DIY SCIENCE FAIR FUN!
ASTRONOMY PROJECT YOUR WAY

Megan Borgert-Spaniol

Super Sandcastle

An Imprint of Abdo Publishing
abdobooks.com

abdobooks.com

Published by Abdo Publishing, a division of ABDO, PO Box 398166, Minneapolis, Minnesota 55439. Copyright © 2024 by Abdo Consulting Group, Inc. International copyrights reserved in all countries. No part of this book may be reproduced in any form without written permission from the publisher. Super SandCastle™ is a trademark and logo of Abdo Publishing.

Printed in the United States of America, North Mankato, Minnesota

102023
012024

THIS BOOK CONTAINS RECYCLED MATERIALS

Design: Aruna Rangarajan, Mighty Media, Inc.
Production: Mighty Media, Inc.
Editor: Liz Salzmann
Cover Photographs: Mighty Media, Inc.; Shutterstock
Interior Photographs: Adobe Stock, pp. 4 (both), 6, 15, 27, 30, 31; Flickr/NASA, p. 5; iStockphoto, p. 26; Mighty Media, Inc., pp. 16 (experiment), 17 (experiment), 18 (experiment), 28 (experiment); Shutterstock, pp. 1, 7, 8, 9 (all), 10, 11 (all), 13, 14 (all), 16 (hand), 17 (balls), 18 (balls), 19, 21 (balls), 22, 24, 25, 28 (moon, rocks), 29 (all)
Design Elements: Shutterstock

Library of Congress Control Number: 2023939284

Publisher's Cataloging-in-Publication Data
Names: Borgert-Spaniol, Megan, author.
Title: Astronomy project your way / by Megan Borgert-Spaniol
Description: Minneapolis, Minnesota : Abdo Publishing, 2024 | Series: DIY science fair fun! | Includes online resources and index.
Identifiers: ISBN 9781098292027 (lib. bdg.) | ISBN 9781098278922 (ebook)
Subjects: LCSH: Do-it-yourself work--Juvenile literature. | Astronomy--Juvenile literature. | Space--Juvenile literature. | Science projects--Juvenile literature. | Science fair projects--Juvenile literature.
Classification: DDC 507.8--dc23

Super SandCastle™ books are created by a team of professional educators, reading specialists, and content developers around five essential components—phonemic awareness, phonics, vocabulary, text comprehension, and fluency—to assist young readers as they develop reading skills and strategies and increase their general knowledge. All books are written, reviewed, and leveled for guided reading, early reading intervention, and Accelerated Reader™ programs for use in shared, guided, and independent reading and writing activities to support a balanced approach to literacy instruction.

CONTENTS

EXPLORE ASTRONOMY 4
BECOME A SCIENTIST! 6
ASK A QUESTION 8
GATHER INFORMATION 10
FORM A HYPOTHESIS 12
PREPARE YOUR LAB 14
EXPERIMENT! 16
RECORD THE RESULTS 20
WRITE A CONCLUSION 22
FURTHER RESEARCH 24
PRESENT YOUR PROJECT 26
KEEP ASKING QUESTIONS 30
GLOSSARY 32

EXPLORE ASTRONOMY

Do you love to look at the moon and stars? Do you wonder if humans could live on Mars? You might enjoy astronomy! Astronomy is the study of outer space. Scientists who study outer space are called astronomers.

Mars is often called the red planet. It gets its color from the rusty iron on its surface.

Astronomers use special antennas to study objects in space.

Astronomers **research** the planets and moons in our solar system. They study faraway stars and **galaxies**. They try to find out how the universe began!

Astronomers use telescopes such as the Hubble Space Telescope to see deep into space.

BECOME A SCIENTIST!

Scientists use a process called the scientific method. Check out the steps on the next page. You will use this method to **design** your own astronomy project!

THE SCIENTIFIC METHOD

1. ASK A QUESTION — What would you like to find out?

2. GATHER INFORMATION — What information do you need to understand your topic?

3. FORM A HYPOTHESIS — What do you think is the answer to your question?

4. EXPERIMENT — How can you test your hypothesis to find out if it is correct?

5. RECORD THE RESULTS — What did you observe in your experiment?

6. WRITE A CONCLUSION — Did your results support your hypothesis?

STEP 1: ASK A QUESTION

What topic do you want to learn about?

Maybe you are interested in the stars. Or maybe you want to know more about the moon's craters. Start asking questions! Write your questions in a notebook so you don't forget them.

Why are some stars brighter than others?

Why do stars appear to move across the sky?

STEP 2: GATHER INFORMATION

Maybe you have a lot of questions. That's great!

Scientists often have many questions they'd like to answer. But for now, choose one to **focus** on. Save the others for **future** projects.

It's time to **research** your **topic**. You can gather **information** from many different sources.

- Read online articles about the topic.
- Read books about the topic.
- Talk to scientists or other experts.

What did you learn about your **topic**? Write it down in your notebook. Then you'll have all the **information** you need in one place.

Craters are bowl-shaped landforms on the moon's surface.

Craters form when space rocks crash into the moon.

Space rocks can be the size of a grain of sand or a small planet!

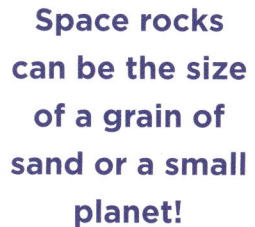

STEP 3: FORM A HYPOTHESIS

After you research your topic, it's time to form a hypothesis.

Your hypothesis is what you believe is the answer to your question. First, revisit your question. Do you want to change it based on what you learned? Then think of a few different hypotheses. Record them all in your notebook.

Question: Why are some craters wider than others?

Hypothesis 1
The width of a crater depends on the weight of the space rock that created it.

Hypothesis 2
The width of a crater depends on the speed of the space rock that created it.

Hypothesis 3
The width of a crater depends on the width of the space rock that created it.

Now, choose which hypothesis makes the most sense based on your **research**.

I think the width of a crater depends on the width of the space rock that created it.

So, the wider a space rock is, the wider the crater it creates will be.

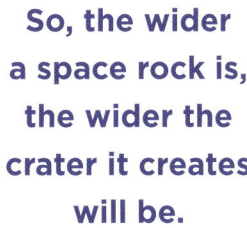

PREPARE YOUR LAB

Get ready to test your hypothesis.

Find an area with a sturdy table or counter to work on. Then gather the supplies you'll need for your science experiment.

SUPPLIES

- large baking dish or tub
- flour
- ruler
- spoon
- cocoa powder
- wire strainer or sifter
- marble
- golf ball
- Wiffle Ball

LAB RULES

All labs have rules that scientists have to follow. Here are some rules for your lab. They will help you stay safe and have fun while doing your experiment!

➜ **Ask an adult** for permission to use the materials and do the experiment.

➜ **Ask for help** with sharp or hot tools.

➜ **Wear goggles** and gloves to protect your eyes and hands.

➜ **Clean up** when you are done and put everything away.

STEP 4 EXPERIMENT!

You've gathered the supplies. You've prepared the lab. It's time to experiment!

1 Pour about 1 inch (2.5 cm) of flour into the baking dish. Use the spoon to even out the flour.

2 Use the sifter to sprinkle a thin layer of cocoa powder over the flour. The cocoa powder will help you see the edges of the craters.

3 Hold a ball 12 inches (30 cm) over the baking dish. Let it drop onto the flour.

4 Gently remove the ball. Measure the width of the crater it created. Write down the crater's width and which ball made the crater.

5 Repeat steps 3 and 4 with the other two balls. Space the ball drops so they each land on a smooth area of flour.

✦ LAB TIP ✦

Try not to touch the flour when you pick up the balls. Using small tongs could help.

Look at the results of your experiment so far. You might be ready to draw a conclusion. But first, consider any other **variables** that might affect your results.

You dropped all three balls from the same height. This means they all fell at the same speed.

The crater widths were different even though the speed did not change.

A Wiffle Ball is wider but lighter than a golf ball. It made a wider crater than the golf ball even though it is lighter.

STEP 5: RECORD THE RESULTS

During experiments, scientists record data and other observations. You wrote down the crater widths created by different balls. Now it's time to record your data to share with others.

Scientists often use tables and graphs. This helps make the results easy for others to understand.

A table organizes **information** in rows and columns.

BALL	BALL WIDTH	CRATER WIDTH
Marble	0.63 inch (1.6 cm)	0.75 inch (1.9 cm)
Golf Ball	1.63 inches (4.1 cm)	1.75 inches (4.45 cm)
Wiffle Ball	2.88 inches (7.3 cm)	2.5 inches (6.35 cm)

A scatter plot graph shows the relationship between two **variables**.

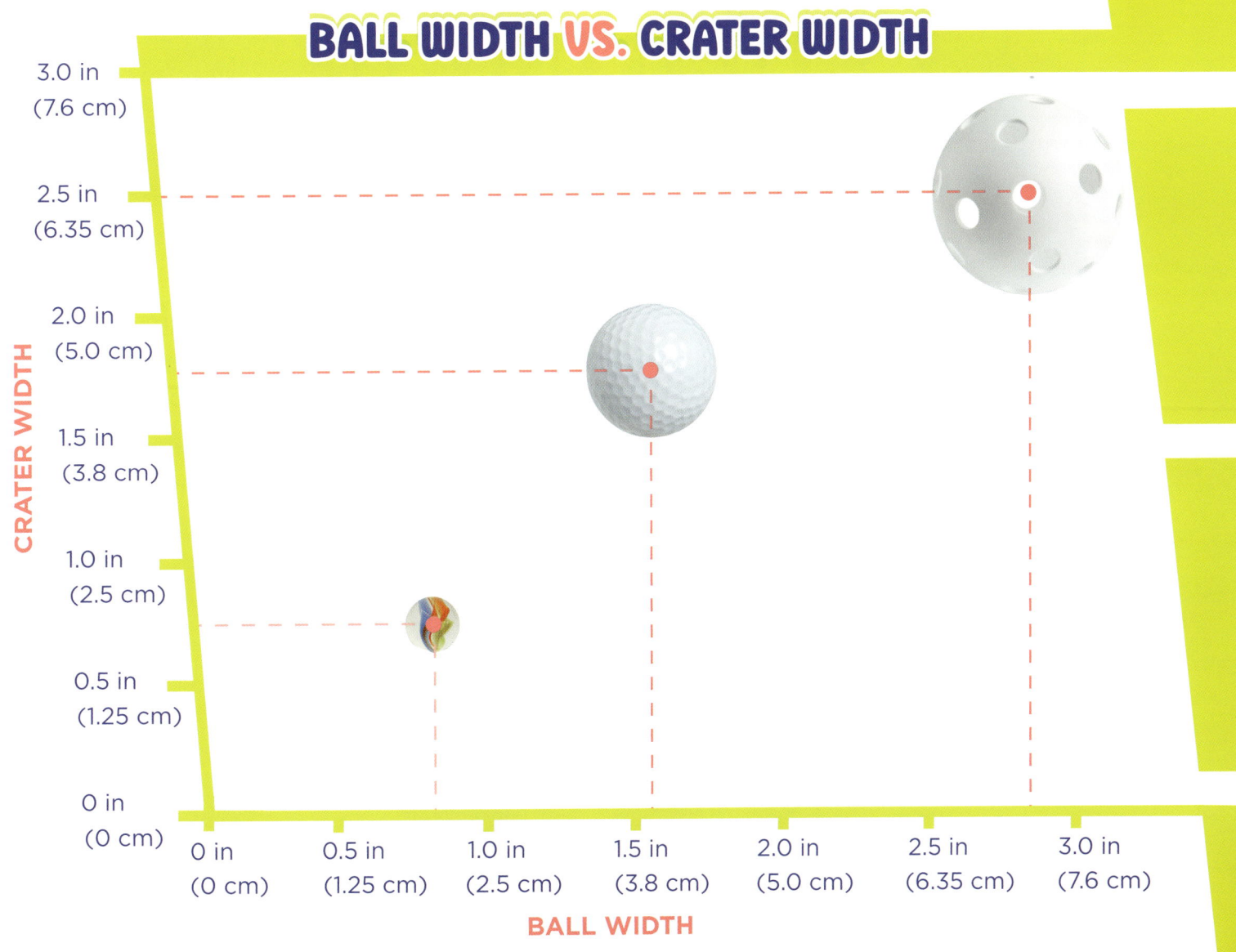

STEP 6
WRITE A CONCLUSION

You recorded your results. Now it's time to write your conclusion. This is a **summary** of your experiment. Your conclusion provides the answer to your original question. It also states whether your results support your hypothesis.

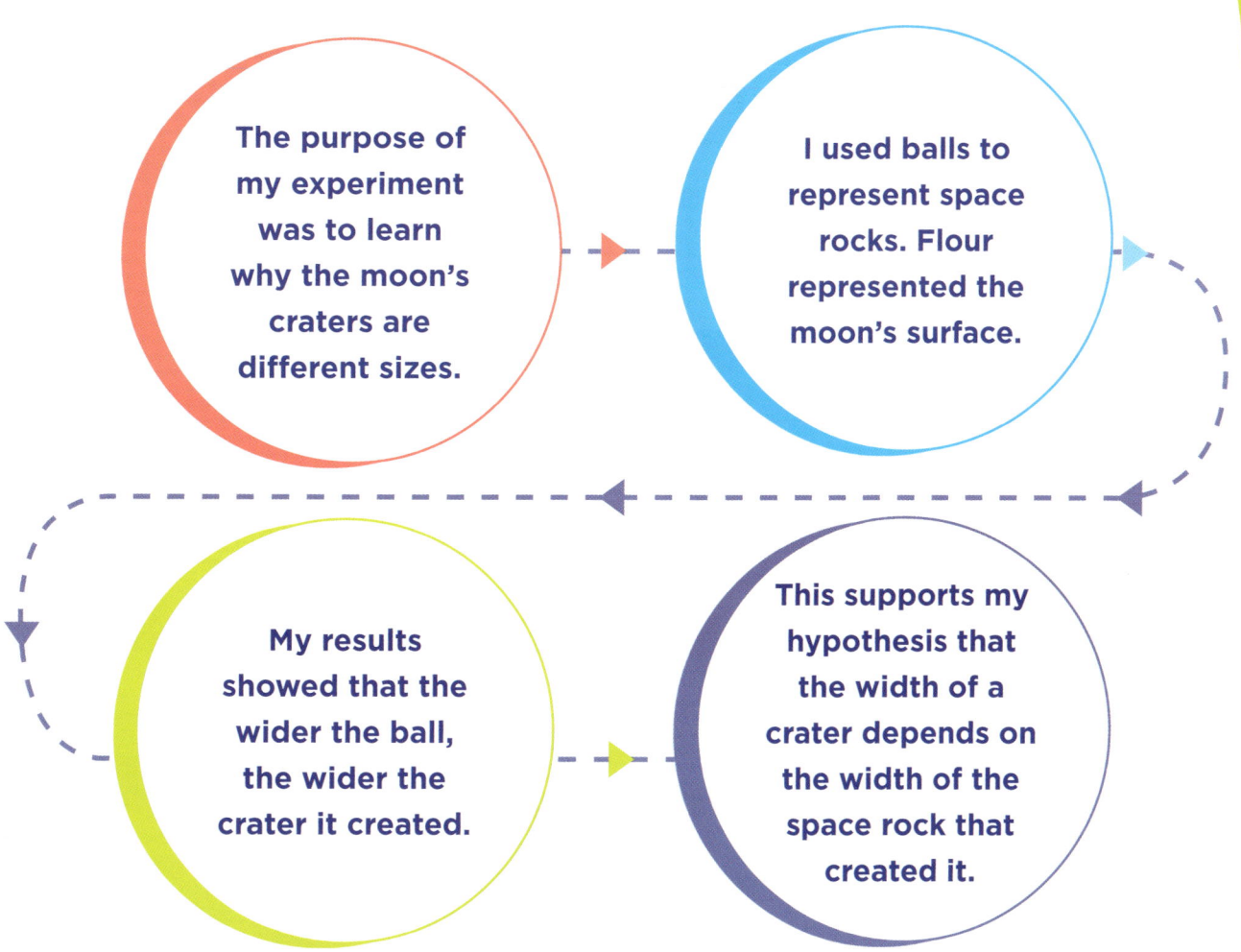

Many scientists find that their hypotheses were wrong. There is nothing wrong with being wrong! It means you did your experiment without **bias** and were surprised by the results. That is another mark of a great scientist!

FURTHER RESEARCH

A conclusion offers an answer to your original question. But it can bring up new questions too! For scientists, the end of one experiment often leads to more **research** and new hypotheses to be tested.

What new questions do you have? What could you research further? Do you have a new hypothesis to test?

Does the speed of a space rock affect the width of the crater?

What affects how deep a crater is?

Does a crater's width change if the space rock hits the moon at an angle?

PRESENT YOUR PROJECT

Young scientists share their **research** at science fairs. Students share what they learned with classmates, teachers, parents, and sometimes judges. It is a chance to show all the work they put into their experiments.

THERE ARE LOTS OF FUN WAYS TO SHARE YOUR NEW KNOWLEDGE!

- Demonstrate or show a video of your experiment.
- Create comics or other drawings to show your project in a fun way.
- Include props, models, or dioramas.

One way to present a project is with a display board. It should show how you followed the scientific method. Turn the page to see a display board of the project in this book!

CRATERS OF THE MOON

QUESTION
Why are some craters wider than others?

RESEARCH
Craters are bowl-shaped landforms on the moon's surface. Craters form when space rocks crash into the moon. Space rocks can be the size of a grain of sand or a small planet!

HYPOTHESIS
I think the width of a crater depends on the width of the space rock that created it.

EXPERIMENT
In my experiment, I dropped different balls onto flour. I recorded the width of the crater each ball created.

Marble

Golf Ball

Wiffle Ball

RESULTS

My results showed that the width of the crater increased with the width of the ball.

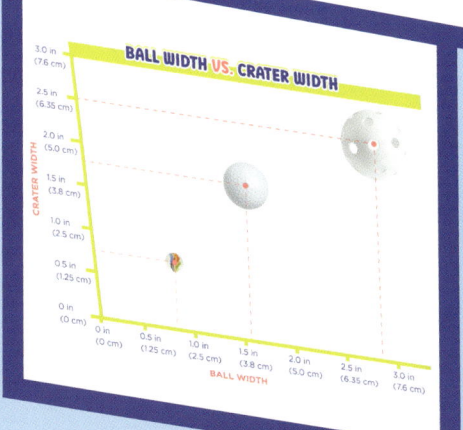

CONCLUSION

The results support my hypothesis that the width of a crater depends on the width of the space rock that created it.

KEEP ASKING QUESTIONS

Your science project is over. You packed away your display. But don't stop asking questions! What might you do differently if you did the project again? What additional **research** could you do? Is there a related **topic** you would like to explore?

Beyond the Science Fair

Be a scientist beyond the science fair! You can use parts of the scientific method to find answers to everyday questions. Maybe you have a hypothesis for why the cake you baked didn't turn out. Maybe you experiment to find the best way to train your dog. One day, you might use science to do big things. Maybe you'll find a new distant **galaxy**! Turn your world into a science fair. What will you discover?

GLOSSARY

bias—showing a preference for one result over another.

design—to plan how something will appear or work.

focus—to concentrate on or pay particular attention to.

future—the time that hasn't happened yet.

galaxy—a very large group of stars, planets, and other objects in space. Earth is in a galaxy called the Milky Way.

information—the facts known about an event or subject.

research—to find out more about something. Also, a study of something to learn new information.

summary—a short statement of the main points.

topic—the main idea or subject.

variable—a factor in a scientific experiment that may change.